I0486751

Christian Quantum

Ross Thompson

Published by Ross Thompson, 2020.

Table of Contents

SUMMA

A story is told of a Professor who asked his student, "What is electricity?" The student replied, "Oh, sir! I'm sure I have learnt what it is, I'm sure I did know, but have forgotten." To which the Professor responded, "How very unfortunate. Only two persons have ever known what electricity is, God the author of nature and yourself. Now one of them has forgotten!"

Quantum Mechanics makes visible God has founded the material realm, His creation, on enigmas and inexplicable microscopic events. None of us will ever be able to sit back in self-satisfaction saying, "now we have all the answers." Not a Mathematician, Scientist or Physicist of the past nor any of the present day, understands the astonishing performance of the atomic and subatomic realm. Many have had pride pricked and been greatly irritated at the discovery of the phenomenal revelation of the secret things of God, uncovered by Quantum Mechanics.

It is exhibited a constant reminder: "...Know that the Lord, He is God; It is He who has made us and not we ourselves. We are His people and the sheep of His pasture..." (Psalm 100:3)

SMALL

I agree with Japanese American Christian Minister, and former scientist, Aiko Hormann. The quantum field of Quantum Mechanics is the border at which the material realm encounters the spiritual realm of the God of the Bible - God the Father, Jesus Christ (His Son) and the Holy Spirit. Mrs Hormann goes further saying Gods soul has an affinity with matter at that level. I'm not sure what she means by that. It does bring to mind Paul speaking about all creation - Acts 17:28. "For in Him (God) we live and move and have our being, as also some of your (the Greeks) own poets have said, 'For we are also His offspring."

We will get too more on those opinions – but first, what is the quantum field and Quantum Mechanics? I think most people these days could give some sort of explanation of the meaning of the terms. The huge number of books, websites, videos, newspaper headlines and magazine articles on these topics, and the way the mystery of the quantum field keeps it resurfacing in the news, makes it just about certain anybody who can read has some idea about the issue.

Still I think a brief explanation should be made. The quantum field is the material realm at the absolute microscopic level. It is the discoveries scientists and physicists have made as they have delved deep into the microscopic layers of matter. It is the realm of atoms, electrons, photons and more than 200 other subatomic particles, and the forces by which they interact and hold together. Particle Physicists, the people who build and use the huge and expensive structures known as particle accelerators and colliders, are confident of increasing the number of detected particles.

Quantum Mechanics can probably be stated as the study of the activity of matters microscopic foundations. How the quantum field provides the base for what we call material reality. It has many surprises. It might be more accurate to say it is all surprises.

"Anyone who is not shocked by quantum theory has not understood it." Niels Bohr, Physicist.

"I think it is safe to say that no one understands Quantum Mechanics... in fact, it is often stated of all the theories proposed in this century, the silliest is quantum theory. Some say that the only thing that quantum theory has going for it, in fact, is that it is unquestionably correct." Richard Feynman, Physicist.

"All modern physics is governed by that magnificent and thoroughly confusing discipline called Quantum Mechanics. It has survived all tests and there is no reason to believe there is any flaw in it......We all know how to use it and apply it to problems; and so we have learned to live with the fact that nobody can understand it." Murray Gell-Mann

"I cannot define the real problem; therefore, I suspect there's no real problem, but I'm not sure there's no real problem." Another quote from Richard Feynman about Quantum Mechanics.

"The bottom line is, the quantum world just doesn't work in the way the world around us works," says author and physicist David Lindley. "We don't really have the concepts to deal with it."

An unexpected outcome for me as I peered into this area of scientific discovery, was the realisation I had skipped over the Bibles multitude of references to matter, and Gods involvement with matter – I had not given them much thought at all.

Very small – very very small - is the thing to keep in mind when considering Quantum Mechanics. Brian Clegg in his book 'The God Effect' (not a Christian book) helps us get it into proportion. Quote: "The simple act of turning on a 100-watt light bulb will produce 100,000,000,000 photons (packets of energy that make up light) every billionth of a second."

A Google search for 'The scale of the subatomic realm,' produced this quote: "Each atom is itself a composite that's one tenth of a billionth of a meter across - sitting on the precipitous edge of a universe between our perceived reality and the quantum world. Electrons hazily occupy much of the atom's empty space. Protons and neutrons cluster in the atom's nucleus, 100,000 times smaller than its atom, and are themselves composed of other stupendously small things: quarks and gluons and more. An electron may have no meaningful property of size but could be thought of as 10 million times smaller than the nucleus of the atom." Did you notice your brain starts to hurt trying to get your mind around that?!

The number of atoms on the head of a pin is ten to the power of eighteen. That is 100,000,000,000,000,0000

One million atoms lined up, side by side would be about the length of this period (full stop) (.)

Quarks, one of the particles in the atom's nucleus, are at least 100 million times smaller than the atom itself.

One breath of air contains ten to the power of twenty-two atoms. That's 10 with 22 zeros.

Here are some facts about your breath and atoms. Any parcel of air such as the litre of air you will exhale in the next breath, will mix with the earth's atmosphere during the next few years. This means of the air you exhaled a few years ago in any breath, about one or two atoms are in every litre of air on earth and inside the lungs of every person on earth. And one or two atoms breathed out by every person on earth in any particular breath, are in your lungs now. One or two atoms from George Washington's first breath, his dying breath and every other breath he ever took are in your lungs now.

So are atoms from each breath of all other people who have ever lived including Jesus. We are all breathing one another. Atoms are ageless. It's only the connection of atoms that change. An oxygen atom might be part of a nerve cell in your brain today, part of an atmospheric water molecule a century from now, and part of a tree many years after that. 'Your' atoms, the ones in your body, have just been borrowed from the air, from the breaths of every person who ever lived, from the Earth – borrowed by the acts of breathing and eating - to be given back sooner or later. (*Art Hobson*)

Genesis two tells us after forming man from the ground God breathed into his nostrils the breath of life and man became a living being. Adam became a spiritually alive person and Gods breath kick started his physical respiration on the sixth day. The plants had been created on the third day and the Sun on the fourth day, ensuring photosynthesis, essential to producing oxygen atoms for Adams breathing, was under way. (photosynthesis in plants takes in the carbon dioxide produced by all breathing organisms and reintroduces oxygen into the atmosphere)

The New Testament records that momentous event when after His resurrection, Jesus breathed on His disciples saying, "Receive the Holy Spirit. If you forgive the sins of any, they are forgiven them; if you retain the sins of any they are retained." (John 20:22) It seems right to conclude the disciples were born again at that time. Men became living beings again, delivered from death - the consequence of trespasses and sins - (Ephes 2:1) in the sense Adam was a living being before the fall. Did Jesus need to breath the atmosphere in His resurrection body? I don't have the answer. Certainly, it was a body beyond the dimensions we know. (John 20:19,26.) We will return to the resurrected Lord further on.

67,500,000,000,000,000,000,000,000,000,000,000,000,000,000,000 is the estimate of oxygen atoms on earth. We sample at most 0.0000000001 percent of all the oxygen atoms on earth if we reach 80 years of age.

GLIMPSE

Take another look at this period (.) How do they see this stuff? At first with the Electron Microscope. In the 1920's physicists discovered that every particle of matter has a certain kind of wave associated with it. These are called psi waves. The microscope shoots a thin stream of electrons which scans the area of the item being looked for (eg) a few atoms. The psi waves of the electrons disturb the psi waves of the atoms. Other instruments then collect and record the patterns made by the electrons. In this manner the first clear image of a string of atoms - thorium atoms - was obtained. Thorium is a naturally occurring weakly radioactive chemical element, under consideration for use in a new generation of nuclear reactors - as an alternative source of fuel for the generation of electricity. Australia has an abundant supply of thorium.

1983 saw the development of the Scanning Tunnelling Microscope, which also uses electrons psi waves but on a much smaller scale. The main feature being a tiny probe the shape of a sharp pencil tip only a few atoms wide. As with the Electron Microscope - patterns are processed by a computer and displayed as a 3D image. The probe tip on this beauty can also pick up individual atoms and move them from place to place. In 1990 IBM scientists picked up 35 atoms of Xenon gas and rearranged them to spell out the name of their laboratory. Xenon atoms do not combine easily with other atoms, making them easier to manipulate.

Xenon is a gas, colourless, dense and odourless. It is found in Earth's atmosphere in trace amounts. Xenon is used in flash lamps and arc lamps, and as a general anaesthetic.

Another attempt at improving the instrument is the atomic force microscope that, expressed in simple terms, works by feeling and recording the surface of atomic structures.

In February of 2019, a team of researchers at Japan's Hokkaido University developed the world's first entanglement – enhanced (particle entanglement is explained further on) microscope, using a technique known as differential interference contrast microscopy. This type of microscope fires two beams of photons at a substance and measures the interference pattern created by the reflected beams - the pattern changes depending on whether they hit a flat or uneven surface. Using entangled photons greatly increases the amount of information the microscope can gather, as measuring one entangled photon gives information about its partner. The Hokkaido team managed to image an engraved "Q" that stood just 17 nanometres (one nanometre is one billionth of a metre) above the background, with unprecedented sharpness.

FORCES

A li Sundermier, in an article in the publication - Symmetry - confirms atoms (everything material is composed of atoms) are mostly empty space. It is not exactly the way a physicist would put it. She means there is not much mass – solid stuff - in an atom. As quoted above, the nucleus (the centre) of an atom is around 100,000 times smaller than the atoms they're housed in. If the nucleus were the size of a peanut, the atom would be around the size of a baseball stadium.

Every type of non-living and living matter in existence is comprised of a core set of just 100 atoms. All elements - liquid, gas and solids - are made up of different combinations of these 100 atoms. The standard record of elements is known as the periodic table. Each naturally occurring element has a unique atomic number (Z) representing the number of protons in its nucleus. [n 2] *A definition of element - is a substance that can't be broken down into a simpler substance.* Most elements have differing numbers of neutrons among different atoms, with these variants being referred to as isotopes. For example, carbon has three naturally occurring isotopes: all its atoms have six protons, and most have six neutrons as well, but about one per cent have seven neutrons, and a very small fraction have eight neutrons. Isotopes are never separated in the periodic table; they are always grouped together under a single element. *Wikipedia*

You may not have noticed the elements found in the ground were not created during the Genesis one, seven-day creation period. Why? Because the ground and water were not created at that time. They were already in existence. (Gen 1:1) God divided the waters and declared "Let the dry ground appear." (Gen 1:6,9)

Nature has five forces by which everything material exists. The first is not acknowledged by Quantum Mechanics. Albert Einstein asked: Can the Quantum Mechanical description of physical reality be considered complete? He concluded that it could not. Given apparently sensible demands on what a description of physical reality must entail, it seemed that something must be missing. "We needed a deeper theory to understand physical reality fully," he said.

The deeper theory is expressed in the following verses from the Bible.

"God who at various times and in various ways spoke in times past to the fathers by the prophets, has in these last days spoken to us by His Son, (Jesus) whom he has appointed heir of all things, through whom also He made the worlds; who being the brightness of His glory and the express image of His person, and upholding all things (some translations say – upholding the universe) by the word of His power, when He had by Himself purged our sins, sat down at the right hand of the Majesty on high...."(Hebrews 1:1-3)

We also have Colossians 1:17,18. "For by Him (Jesus) all things were created that are in heaven and on earth, visible and invisible, whether thrones or dominions or principalities or powers. All things were created through Him and for Him. And He is before all things, and in Him all things consist."

Paul speaking of the world and people in general "God who made the world and everything in it, since He is Lord of heaven and earth, does not dwell in temples made with hands.........for in Him we live and move and have our being..." (Acts17:24,28)

Further on we will see God has established automatic forces in matter that keep everything operating. Somehow behind all that Jesus upholds all things by the word of His power. The Hebrews verses seem to imply a continuous activity. We are seeing the meeting of the material realm with Gods supernatural realm, where God who is a Spirit, (John 4:24) acts upon His material creation.

Non-Christian physicist John A Wheeler felt compelled to admit the discoveries of the quantum field, "Convey the idea that every item of the physical world has at bottom - at a very deep bottom, in most instances - an immaterial source and explanation."

Upholding all things by the word of His power is something specific to Jesus. If the translations are correct it is something only, He does. He would then still have been upholding all things by the word of His power, in some way, as He walked on the earth. That would have included the atoms of His own human body in which He was to be a sacrifice for our sins.

Physics and Quantum Mechanics have brought to light four forces in nature.

Gravity: Quantum Mechanics has never been able to fully explain gravity. It is the force that keeps us, and everything else stuck to the earth, preventing us from flying off into space. Let's hope it keeps on keeping on!! Gravity keeps the Sun together and in place, and the planets orbiting that hot ball of gas. Albert Einstein, the most famous of physicists, theorised gravity acts on energy, light, radiation, space and time. Experiments proved him correct. Whatever gravity is it permeates all of space. The other forces are mediated by particles, they can be quantized, meaning they could be expressed as individual particles and have noncontinuous values.

Gravity doesn't seem to be like that. The best theory so far moots a hypothetical massless particle called a graviton. It has not got beyond theory. Nobody has found a graviton yet. Fame and fortune are certain for anyone who does. It is not looking good they will ever be found. If they exist, they rarely interact with matter and present-day Particle Detectors would find them to be invisible against what is called, background noise.

Because gravitons haven't been observed yet, gravity has resisted attempts to understand it in the way we understand other forces – as an exchange of particles. Some physicists, notably Theodor Kaluza and Oskar Klein, posited that gravity may be operating as a particle in extra dimensions beyond the three of space (length, width, and height) and one of time (duration) we are familiar with, but whether that is true is still unknown.

The Lord did not mention gravity specifically to Job when He asked, "Where were you when I laid the foundations of the earth? Tell me if you have understanding. Who determined its measurements? Surely you know! Or who stretched the line upon it? To what were its foundations fastened? Or who laid its cornerstone. When the morning stars sang together. And all the sons of God shouted for joy?" (Job 38:4-7) We could be reading about the invention of gravity in that passage. Whenever it was it seems to have been a big event.

Since gravities main theatre of operation is space, what does the Bible say about space. Chuck Missler says, "space is not an empty vacuum: It can be torn (Isaiah 64:1); worn out like a garment (Psalm 102:25); shaken (Hebrews 12:26, Haggai 2:6, Isaiah 13:13); burnt up (2 Peter 3:12); split apart like a scroll (Revelation 6:14); rolled like a mantle (Hebrews 1:12) or a scroll (Isaiah 34:4).

God, who alone stretches out the heavens (Job 9:8); Stretching out heaven like a tent curtain (Ps 104:2); Who stretches out the heavens like a curtain, and spreads them out like a tent to dwell in (Isa 40:22); He has stretched out the heavens (Jer 10:12); The Lord who stretches out the heavens (Zech 12:1).

Stretching the Heavens: 2 Sam 22:10; Job 9:8, 26:7, 37:18; Psalm 18:9, 104:2, 144:5, 40:22, Isaiah 42:5, 44:24, 45:12, 48:13, 51:13; Jeremiah 10:12, 51:15; Ezekiel 1:22; Zechariah 12:1.

Electromagnetic force: All matter consists of equal numbers of positive and negative electric charges, which don't repel. That is why ordinarily, all matter is electrically neutral. Which is why we are not much aware of this field. Our bodies are neutral. But slide your socked feet across a polyester carpet then reach out for a metallic doorknob. The spark and zapping sound is your body releasing the excess electrical charge.

The basic unit of negative electric charge is the electron. It is an infinitesimally small sub-atomic particle. You would need to collect 1000 trillion trillion of them to reach a gram in weight. Electrons spin on their own axis. It is not understood why they spin. Science can change the direction of the spin but cannot stop it. The electric charge of the electron spins with it making them magnetic. Electrons move at around 2,200 kilometres per second within the atom – slightly less than 1% of the speed of light.

The basic unit of positive electric charge is the proton. (These are the names we give to these particles – I sometimes wonder if God has His own names for them) Once again they are very small, but a protons mass is 2000 times that of an electron. Protons along with certain neutral particles called neutrons, account for the mass of matter. All atoms consist of equal numbers of protons and electrons. There is an exact equality between numbers of protons and electrons in the Universe. One percent of difference between the negative charges and positive charges inside just one ounce of ordinary matter, would see it torn asunder by a force equal to the total weight of the earth. The electromagnetic force is 'there'. It can never be created or destroyed.

Protons and neutrons are in the nucleus/centre of the atom, which is one ten trillionth of a centimetre across. Electrons circle around and outside the nucleus but still within the atom. Electrons are held in place by the positively charged nucleus. Positive and negative charges attract each other. Therefore, the force that accounts for the structure of the atom is the electromagnetic force. This electrical attraction between nucleus and electron is 10,000 trillion trillion times stronger than the gravitational force between them.

The strong nuclear force: The electromagnetic force keeps the atom together in that it keeps electrons orbiting in it. The strong nuclear force keeps the nucleus of the atom together. Though it is not completely understood, Quantum Mechanics has established how the strong nuclear force operates inside the nucleus, and how it keeps the protons and neutrons together. When I first read of the strong nuclear force, I thought, 'well this is Jesus upholding the Universe by the word of His power.' More reading revealed the nuclear force is an established force in nature, as a part of all material things. Its power is seen in the detestable creation of that abhorrence, the atom bomb.

It is an extremely strong force because it is required to keep the half of the mass of the atom's nucleus, composed of protons, together. They naturally repel each other, and more so when they are very close together. The nucleus would be blown apart without the action of the strong nuclear force. The strong nuclear force is many times stronger than the electromagnetic force. It is though, very short range being confined mostly to the atom nucleus. The electromagnetic and the strong nuclear force combine to make matter very stable.

Playing around, to use a colloquial term, with the strong nuclear force, created the terrible devastation of the atom and hydrogen bombs and the release of radiation - dangerous to all living things. It also established the international standoff we live with today regarding nuclear armament.

I was interested to read, Ernest Rutherford, the scientist who split the atom which started the trek to the atom bomb, did not think it had any potential as a source of power. He is quoted as saying, "We might in these processes obtain much more energy than the proton supplied, but on the average, we could not expect to obtain energy in this way. It was a very poor and inefficient way of producing energy, and anyone who looked for a source of power in the transformation of the atoms was talking moonshine. But the subject was scientifically interesting because it gave insight into the atoms." He died before the production of the A-Bomb.

The processes that resulted in the bomb have also been harnessed for power production and in medicine and industry. In hindsight and bitter experience, nuclear reactors for power have proven to be an expensive and perilous concept that might have been better left alone. Especially since now, only a few decades on, they are becoming obsolete in favour of alternative and safer power sources such as wind farms, solar power, hydrogen and others.

Nuclear fission in the reactor is a volatile and unpredictable event. Enriched uranium (enriching is the process that separates fissionable light U235 uranium from non-fissionable heavy U238 uranium) releases neutrons that collide with and split atoms of another uranium product. More neutrons are released, and more atoms split producing a chain reaction. The process causes the original uranium product to become extremely hot. This heating and the chain reaction are controlled by the insertion of rods which are an amalgam of neutron absorbing materials – such as silver, boron, cadmium and cobalt. Raising or lowering the rods controls the number of neutrons reaching the second uranium product – thereby controlling the chain reaction. The whole process is fraught with danger. The threat of a nuclear explosion always exists if a system fault or other event (eg earthquake) allows the chain reaction to get out of control.

The period between 1957 to 2011 saw twenty-six major accidents in nuclear reactors. People were killed and multiplied millions of dollars lost. Harm from radioactivity released into the surrounding atmosphere can only be guessed at. The operating system in a reactor is a delicate balance needing only the slightest malfunction or human error to produce disaster. Intensifying the hazard is the brief time span reactor staff have to stave off a calamity. In a chain reaction 1,500,000,000,000,000,000 atoms are split in a fraction of a second. The central uranium mass requires constant cooling and will soar to unimaginable temperatures without it. A by-product of some reactors is polonium – not the most poisonous element known to man, but close – with a lethal life of 400,000 years. The sodium solution used in cooling will ignite if it contacts air. The many reactors decommissioned because of accident are sealed off and guarded day and night for at least sixty years. Safely disposing of reactor radioactive waste is a constant problem. It is not surprising at 2016 - 40 countries made the decision to drastically reduce reliance on nuclear reactors for power supply.

Weak nuclear interaction: This least understood of the forces is still very much stronger than gravitational force. But weak compared to the electromagnetic force. It occasionally combines protons, electrons and neutrinos to form neutrons. Neutrinos are everywhere in the Universe. They travel at the speed of light and are electrically neutral allowing them to pass right through the Earth, you and me, and planets and stars without being hindered. The weak interaction is active in the area of the extremely small. Hence the mystery surrounding it. A branch of Quantum Mechanics known as Particle Physics is focused on the investigation of the unimaginably small inside the atoms nucleus and is responsible for detecting the 200 other sub-atomic particles mentioned at the beginning of this book.

Each of these forces is dominant in a different domain. The strong nuclear force rules in the atom nucleus, but nowhere else in the rest of the atom. The electromagnetic force prevails between the nucleus and electrons of the atom and between atoms and molecules. Molecules are the next step up in matter from the atom. The gravitational force governs the motions of the stars but is inconsequential inside the atom.

HARMONY

A Physicist who is not a Christian, who does not have the Holy Spirit within him, is at a great disadvantage. If he hopes to discover the secrets of life through his profession, he will be disappointed, for two reasons. He has no spiritual insight and secondly the area of his study, the material realm, is not in its optimal condition. The temporal realm as the Bible calls it is decaying. Psalm 102 v25-27 "Of old you (God) laid the foundation of the earth, and the heavens are the work of your hands. They will perish but you will endure. Yes, they will all grow old like a garment. Like a cloak you will change them, and they will be changed." 2Peter 3:10,12,13 "But the day of the Lord shall come like a thief in the night, in which the heavens will pass away with a great noise, and the elements will melt with fervent heat; both the earth and the works in it shall be burned up........the day of the Lord, because of which the heavens will be dissolved, being on fire, and the elements will melt with fervent heat. Nevertheless we, according to his promise, look for new heavens and a new earth in which righteousness dwells."

The Onkelos Bible translation calls the original creation a unified order. When it should be in harmony with the eternal Spirit who is God, it is, because of sin, separated, temporary and doomed to destruction. Our bodies will return to dust as will everything the present quantum realm supports.

Even we who are Christians who are back on the road to the perfecting of all things, have little understanding of a material realm perfectly in harmony with God the Spirit. We see it in the first chapters of Genesis. There, God the Father, Jesus and the Holy Spirit (Gen 1:26) are in the act of creating man, the earth, light, the sky, nature, the sea, the planets, the sun, the moon, and all other creatures. A material realm perfectly aligned with its Spiritual creator is revealed in various verses.

"And the Lord God formed man out of the dust of the ground, and breathed into his nostrils the breath of life; and man became a living being" (Gen 2:7)

"The Lord God planted a garden eastward in Eden" (Gen 2:8)

"And they (Adam and Eve) heard the sound of the Lord God walking in the garden in the cool of the day... (Gen 3:8)

"And God said, 'let there be light' and there was light." (Gen 1:3)

"Then God saw everything that He had made and indeed it was very good." (Gen 1:31)

The blessing is - the book in which we read these verses has been given to us as a guide to get us back to our spiritual Creator. The scientist, physicist or anyone else who will take the time to investigate the Bible will find that God has provided a path back to Himself for any and all who will receive it. Christ died to take the penalty for sin – the cause of the decay of the temporal realm – and to open the way for an immediate return to relationship with God the Spirit, our creator.

"God...who has saved us and called us with a holy calling, not according to our works, but according to His own purpose and grace which was given us in Christ Jesus before time began, but has now been revealed by the appearing of our Saviour Jesus Christ, who has abolished death and brought life and immortality to light through the gospel...." (2Tim 1:9,10)

LAYMAN

I am no Physicist. Mathematics is a mystery to me. I very quickly reach for a calculator for basic arithmetic. I don't seem to have a mind for numbers. When I read books such as Brian Clegg's 'The God Effect' – a book written to give the ordinary person some understanding of Physics - a large portion of the book eludes me. I have understated that actually – a lot of the time I don't have a clue what he is talking about!! I find I go through a process. I'll read what he says on something a few times. If I'm not getting it, I'll go to the dictionary to find the exact meaning of particular words he has used. Then I will Google search (for example) 'Quantum entanglement explained simply'. As a last resort I may Google, 'teaching physics to children.'

I found some comfort in the discovery physicists don't always understand each other or get it right. In 1900, the British physicist Lord Kelvin, who gave science the Kelvin temperature scale, is said to have pronounced: "There is nothing new to be discovered in physics now." He was a Christian. One scientist was asked what he thought of his colleague John A Wheeler's ideas? "I have not the slightest idea what he means," was the reply. Asked the same question, another replied. "It's interesting but I don't really understand it" Which illustrates the fact there seem to be physicists and physicists. Some seem not bothered about understanding the enigma of Quantum Mechanics, simply getting on with the job of making it work in the real world. Theoretical/ philosophical physicists like John A Wheeler spend much time in thought experiments looking for the why's and wherefores. Albert Einstein declared imagination was an important tool in physics.

Now that you have stopped laughing, (thank God I seem to make some progress with reading and writing!!) I would like to say, in a way my ignorance doesn't matter, regarding getting a superficial knowledge of Quantum Mechanics. All the discoveries of Quantum Mechanics are the result of multiple experiments by a great many Scientists. Since the Quantum field came to light around the late 1800's early 1900's, each discovery has been checked and rechecked by other scientists in the field.

Often subsequent experiments will add something new to the discovery leading to more experiments to qualify that discovery and so the process goes on. When reading books on Physics you notice quickly (and I suppose it is the same with all knowledge) that a certain scientist reads a paper by a colleague which stimulates his thinking. A new angle comes to mind from that reading, an experiment follows, and a new discovery comes to light. A colleague reads of his results, has a new thought and on it goes.

My point is that all Quantum Mechanical knowledge has been confirmed many times over and can be relied upon as factual without having to understand the complicated mathematics, experiments or instrumentality that enabled the discovery. People like you and I can read the simple version of the facts and gain a reasonable understanding. I have an article from a Science website entitled, 'The Strong Force is what's holding the Universe together'. That's an explanation I can understand. The discovery of our existence being reliant upon forces makes one hope whoever is in charge of the Universe forces switchboard, does not hit the off button by mistake!! I'm joking. There will be a new heaven and a new earth, and the old heaven and earth will pass away. But I am sure God will be overseeing the whole process. We have the end of the story and it is a happy ending. (2Pet 3:13) (Rev 21:1)

Armed with my superficial but factual knowledge of Quantum Mechanics and knowing the God of the Bible is the creator of this world we live in, including the bodies and souls and life we possess – I began to think of some instances in the New Testament in particular that show

God doing things with His creation that have a remarkable similarity to the discoveries of Quantum Mechanics. In some instances, God perfectly executes an exercise the physicists have envisaged, but failed to execute and have completely given up on.

Physicists have toyed with idea of moving the atoms of which an object is comprised, to another location. These were experiments with quantum particles based on quantum entanglement, a strange fact of the sub-atomic realm we will get too later. These brief and failed experiments were an attempt to replicate the 'beam me up Scotty' transporting in the Star Trek series.

The best they have been able to do is add a third particle to the entanglement (discussed later) of two sub-atomic particles and thereby get a sort of cloning of one of the particles. Brian Clegg discusses these experiments. Quote: "There is a long way to go down a path that could see the transportation of a living creature however simple....and the leap from there to large scale life such as a human being is even greater and probably will never be practical. To be transported, your human body would have to lose its quantum uniqueness. It would involve nothing less than total disintegration." *Unquote.* He then goes on to mention the problem of the soul and mind being transported with the body. For scientists it remains an interesting idea probably motivated by wide-spread knowledge of the Star Trek television series, which incidentally sidesteps the problem of transporting mind and soul.

God on the other hand has no need of experiments. He does what He wants with His creation. Let's look at the book of Acts chapter 8:37-40 in the New Testament.

"Then Philip said, 'if you believe with all your heart, you may.' And he answered and said, 'I believe that Jesus Christ is the Son of God.' So, he commanded the chariot to stand still. And both Philip and the eunuch went down into the water, and he baptised him. Now when they came up out of the water, the Spirit of the Lord caught Philip away, so that the eunuch saw him no more; and he went on his way rejoicing.

But Philip was found (some translations say appeared) at Azotas. And passing through, he preached in all the cities till he came to Caesarea."

The Spirit of the Lord picked Philip up and transported him body, soul and spirit approximately 17 miles away to the town of Azotas. He, being the dedicated disciple he was, seems not to have complained to God that he had not had time to pack. He continued 55 miles to Caesarea preaching the gospel as he went and no doubt trusting the Lord to supply whatever need he had.

We would have to ask Philip about his experience to know exactly what God did with him. Did he just disappear from where he was and suddenly find himself standing in Azotas? In that case I can envisage him having to ask a local, "What's the name of this town?" Or did he get a panoramic view of 17 miles of desert as he was carried in the Spirit over it. Either way in a time well before aircraft, it was no doubt a faith building, life changing experience for Philip.

John G Lake testifies in one of his sermons that during a church service in South Africa he was kneeling praying at the pulpit and was lifted out of his body – the congregation could see him continuing to kneel – and taken to a Psychiatric hospital in Wales. He watched the earth below him as he 'flew' over it. He crossed the coast of Wales, continuing to the town at which the hospital was located. He 'landed' in the room where a certain tormented lady was. He prayed for her and 'flew' back. A few weeks later a letter to a relative in his church confirmed the lady had miraculously recovered and been released from the hospital to lead a normal life.

God did the same with Ezekiel three times. Ezekiel chapters eight, eleven and forty. "He stretched out the form of a hand, and took me by a lock of my hair: and the Spirit lifted me up between earth and heaven; and brought me in visions of God to Jerusalem..." (Ezekiel 8:3) Surely the disservice the captives had done to themselves in rejecting their relationship with God must have occurred to Ezekiel as he experienced the ease with which God took him back to Jerusalem. The difference here is it happened too Ezekiel as part of a vision.

I remember the African Prophet, Shepherd Bushiri, a man from Malawi, who has attendances of 80,000 at his meetings in the Pretoria Show grounds, South Africa - telling of a time he was driven to his home by 'his protocol' the African term for his assisting team. The key to his home had been misplaced and the search was on as he waited in his car. Finally, the key was found, the door unlocked. and the Prophet was found sitting inside the house. Clearly God has no difficulty manipulating the matter and persons He created.

Israeli Physicist Samuel Braunstein working on a hypothetical that were it possible for scientists to transport a human being Star Trek style, from one place to another, estimated how much information would have to be transmitted to perform such a feat. If a body so transported, had just one molecule a millimetre out of place, the result would be disaster for the healthy operation of the body. Even if it were possible somehow, Mr Braunstein calculated a billion trillion computer hard drives would be required, or a bundle of CD Rom disks that would take up more space than the moon. Time involved would be 100 million centuries to transmit the data for one human body from one place to another. Physicist Braunstein concluded, "It would be easier to walk!"

Returning to the subject of the Lords' resurrection body. Luke twenty-four tells us Jesus drew near to the two disciples walking on the Emmaus road. He talks with them then disappears from their sight. "Now it came to pass, as He sat at the table with them, that He took bread and broke it and gave it to them. Then their eyes were opened, and they knew Him; and He vanished from their sight." (Luke 24:30,31). They raced to report to the twelve and those gathered with them and in the middle of their testimony Jesus suddenly stood in the midst of them. He bids them touch Him and asks for food which He eats. He clearly explains He is not a spirit but has flesh and bones. At another time, recorded in John chapter twenty, Jesus appears to the disciples while the doors were shut, then again eight days later. "And after eight days His disciples were again inside, and Thomas with them. Jesus came, the doors being shut, and stood in the midst said, "Peace to you." (John 20:26)

In his book 'The Physics of Immortality,' Chuck Missler discusses these appearances by Jesus. Quote: Yet while Jesus was physical and tangible and quite alive. He was also able to do some remarkable things in His resurrection body. Specifically, He was able to disappear and reappear at will......We also recognise that Jesus in His resurrection body could walk through walls. He could enter or leave rooms without passing through the walls, floor or ceiling, which suggests that He existed in more than our four space time dimensions....Jesus was raised in a new glorified body that was no longer subject to time and space. It was a body that could step in and out of our four-dimensional, space time domain."

Adding to the mystery John 20:27 tells us, Jesus invited Thomas to reach out and put his hand in the wound in His side. His wounds were still open, they were not bleeding and presumably He had no pain!

As I was writing this I thought, "Where was Jesus between these visitations?" Then I remembered John chapter twenty. Mary stands weeping at the empty tomb and Jesus appears to her. She thinks He is the gardener, but Jesus declares Himself to her. "Jesus said to her, 'Do not cling to me for I have not yet ascended to my Father, but go to my brethren and say to them, I am ascending to My Father and your Father, and to My God and your God." (John 20:17) We have seen later in John twenty Jesus invited the disciples to touch Him. Therefore, He had ascended to the Father and returned. I think it is reasonable to assume He was returning to heaven between these visits.

Chuck Missler suggests John was thinking of these appearances of Jesus when He said, "Beloved now are we the sons of God; and it does not yet appear what we shall be, but we know that when He shall appear, we shall be like Him for we shall see Him as He is."(1John 3:2)

He goes on to say, "The full resurrection body of Jesus is something we will only be able to understand and appreciate when we ourselves have the same body. I don't think He has only five or six dimensions. I believe He exists in the full ten or eleven dimensions (part of Quantum Mechanics current cutting- edge theory) that make up the completeness of reality." For me (the author) all that can be summed up nicely in that Christian song title 'Heavens going to be a blast!'

Chuck is obviously a fan of string theory – small vibrating strings undergirding all microscopic matter, needing ten or eleven extra dimensions as a crucial part of the theory. Where it's at in Quantum Mechanics presently. Don't hold your breath regarding string theory though. The emphasis is on 'theory'. It's not a matter of locating some small particles and having a closer look to find some vibrating little strings. The mathematics involved puts the size of the strings at a million billion times smaller than the smallest discovered particle. I feel inclined to stop right there! The current process of finding newer and smaller particles, is the use of particle accelerators/colliders. They bring about massive collisions between particles to break them up into smaller particles, which are then kept track of by very expensive instrumentation. Physicists agree there will never be enough money nor the technology to build colliders of the size necessary to look for the strings. They are hoping theorists will mathematically prove their existence.

The mathematic equations involved revealed extra dimensions beyond the three we are familiar with would have to exist.

The idea has surfaced that there may be two kinds of spatial dimension. The dimensions we live in and others extremely small at the super-microscopic level. The strings being so small would vibrate in all dimensions. Vibration of known particles give indication of their mass, electric charge etc. If the geometry of the extra dimensions was known predictions might be possible regarding the strings. Eugenio Calabi and Shing-Tung Yau have come up with the extra six-dimension geometric shapes using algebraic geometry – but too date string theory remains fastened to the notice board of theory.

STRANGENESS

The quintessence of Quantum Mechanics is strangeness - popularly referred to as quantum weirdness. Here are some descriptions of the counter-intuitive properties of the quantum field that continue to astound physicists.

Nonlocality: David Bohm, from the University of London at the Lawrence Radiation Laboratory, noticed that in plasmas {*A plasma is a hot ionized gas consisting of approximately equal numbers of positively charged ions and negatively charged electrons. The characteristics of plasmas are significantly different from those of ordinary neutral gases so that plasmas are considered a distinct state of matter from liquids, solids and gas. An ion is a charged atom or molecule. It is charged because the number of electrons, do not equal the number of protons in the atom or molecule. When an atom is attracted to another atom because it has an unequal number of electrons and protons, the atom is called an ion}* particles stopped behaving like individuals and started behaving as if they were part of a larger and interconnected whole. At Princeton, in 1947, he continued his work in the behaviour of oceans of particles, noting their highly organized overall effects and behaving as if they knew what each of the untold trillions of individual particles were doing. Bohm's interpretation of quantum physics indicated that at the subquantum level location ceased to exist. All points in space become equal to all other points in space, and it was meaningless to speak of anything as being separate from anything else. Physicists call this property nonlocality.

Quote: The world of Quantum Mechanics is non-causal and non-deterministic. Nothing is definitively real; one cannot say anything about what things are doing when we are not looking at them. Reality is non-local: distant particles seem to be inseparably connected into some indivisible whole. Although they can sometimes behave as if they were a compact little particle, physicists have found that they literally possess no dimension. Everything is probabilistic in some strange way.

Quantum entanglement: *Quote*: If two particles physically react with each other quantum theory says their psi fields/wave component usually become intimately connected. The two particles then become a single quantum system with a single shared psi/wave field. The particles are then said to be entangled. From then on everything that happens to one of those particles instantly happens to the other even if they are a million miles apart. They could be in different galaxies, yet quantum theory predicts the same results from many experiments. They behave as though they are one single object in different places. Researches have even taken to calling entangled particles a two–particle.

Quote: Quantum theory predicts that entangled particles exhibit correlations only by the existence of real non-locality. That is instantaneous and distant connections between the particles. That is, a physical change in one particle, such as might be caused by a measurement on that particle, causes instantaneous physical changes in all other particles that are entangled with that particle no matter how far away those other particles may be. Thus, even though they are distant from each other, entangled particles behave like a single object.

In space/time the speed limit for everything is the speed of light. 186,000 mi/sec or 300,000 ks/sec. Like the four forces it is a law of space/time. Entangled particles it seems, missed that lecture while at Uni. The particles themselves don't travel faster than light, but the information or connection between them does. A change to one means instantaneous change to the other no matter how far apart. They thumb their noses at the speed of light!!

This quantum strangeness, entanglement, always reminds me of a Christian strangeness, reasonably common, well documented and having many witnesses. So far, I have not been able to find anything in the Bible that refers to it. The Bible College I attended many years ago was part of a Missionary Society to the South Pacific. A large Church in Vanuatu was the first church established. The leader of the organisation, Apostle N Thomas, regularly held well publicised meetings in Vanuatu. People came from all the islands. Before one meeting an older native man walking to the gathering from his village, through 20-30 kilometres of bush, was pleasantly surprised to have Mr Thomas join him in his walk, and the two had some good conversation. I can't remember if Apostle Thomas had arrived in the country at the time, but he was certainly nowhere near the locality of the old man. He had no knowledge of the event until it was related to him.

Sid Roth of 'It's Supernatural' spoke on camera of a time he visited a town somewhere in Texas. Once again, I can't remember if Sid had been to the town previously. He certainly has never stayed at the Hotel he checked into. To his surprise the Lady proprietor expressed her happiness to see him back for another visit. He earned a peculiar look at his insistence this was his first visit.

John Crowder of Sons of Thunder Ministries had decided not to accompany a team to an overseas location for meetings. The outreach got underway and at one of the meetings the group leader noticed John casually leaning against the wall listening to the proceedings. His friend was surprised but figured John had changed his mind and decided to join them. Later it was realised John had not changed his plans and knew nothing of the incident.

African Prophet, Uebert Angel, gives these testimonies of experiences in his ministry. I have quoted them at length from his book 'Hearing Gods Voice' because they fit this discussion.

"A few years ago, while I was in my then home at Ashton Underlyne, Manchester, at about 12:30 am one morning. I had just gone in there to pray by myself …the whole house was quiet …everything was in lockdown and I was the only one still up. Just a few minutes into my quiet time the Lord Jesus walked into the room….I saw Him and heard His footsteps and He took me outside the house in that vision, through the wall of the house and outside to the back garden and began to talk to me about different things He wanted us to do in the Ministry. I thought to myself, when the Lord finishes instructing me, I will be just exactly where I was in the house. When He was about to leave, I walked with Him as we continued to talk until we were a few yards from the house, and He disappeared right before my eyes. To my complete and utter surprise, even though the vision had ended I was still outside…. I mean my physical body got out of a locked house in a vision through a wall. I was locked out and had to call my wife to unlock the door to let me inside. The vision had affected my physical being in the natural. At that level of mara vision you can be translated from one location to another in an instant." *Unquote*

At another time He was ministering in Harare, Zimbabwe. The meeting was live by media to the London branch of the ministry. While preaching he began to have a vision seeing himself preaching at the gathering in London. The media phone rang, and a hysterical voice began telling him as they were watching him by media, he appeared in London preaching. Again, he calls this a mara vision where actual physical translation to another location happens during the vision. That may be an explanation of Ezekiel's experience previously mentioned.

Recently, scientists in China reported a new record for entanglement. They used a satellite to entangle six million pairs of photons (light) The satellite beamed the photons to the ground, sending one of each pair to one of two labs. The labs were twelve hundred kilometres (750 miles) apart. And each pair of particles remained entangled, the researchers showed, when they measured (caused a change) one of a pair the other one was affected immediately. They published those findings in the peer reviewed journal Science.

More testimony from Prophet Angel, "At one time I attended a conference where they had a video recording of me doing a healing service in their Church which I had never visited. When I arrived on a Saturday for my speaking engagement the people were thanking me for the service I had apparently conducted on the previous day, yet on that day I was in my Church at a prayer meeting. Scores had been healed. This happened in 2009 in Rotherham here in the United Kingdom." *Unquote*

The Hebrew word Mara is one of several words used to describe visions in the Bible. Prophet Angel says a mara vision makes physical changes in the person experiencing it. His books can be purchased at online suppliers or his website.

Indeterminacy: The meaning of its opposite will help to grasp this quantum strangeness. Determinism is the philosophical belief that all events are determined completely by previously existing causes. One characteristic feature of quantum theory is indeterminacy. The essence of this is that identical physical starting points lead to different outcomes. When we say the position of a microscopic particle is indeterminate, we do not mean simply that we lack knowledge of its position. Rather we mean the particle has no definite position. It is as though a baseball could be either white or spherical but not both at once. Niels Bohr, Physicist

The uncertainty principle is one of the most famous ideas in physics. It tells us that there is a fuzziness in nature, a fundamental limit to what we can know about the behaviour of quantum particles and, therefore, the smallest scales of nature. Of these scales, the most we can hope for is to calculate probabilities for where things are and how they will behave. Unlike Isaac Newton's clockwork universe, where everything follows clear-cut laws on how to move, and prediction is easy if you know the starting conditions, the uncertainty principle enshrines a level of fuzziness into quantum theory.

One way to think about the uncertainty principle is as an extension of how we see and measure things in the everyday world. You can read these words because particles of light, photons, have bounced off the screen or paper and reached your eyes. Each photon on that path carries with it some information about the surface it has bounced from, at the speed of light. Seeing a sub-atomic particle, such as an electron, is not so simple. You might similarly bounce a photon off it and then hope to detect that photon with an instrument. But chances are that the photon will impart some momentum to the electron as it hits it and change the path of the particle you are trying to measure. Or else, given that quantum particles often move so fast, the electron may no longer be in the place it was when the photon originally bounced off it. Either way, your observation of either position or momentum will be inaccurate. *Alok Jha*

The observer effect: The strangest of all. The most astounding mystery of the quantum. When an electron is observed it is a particle, but between observations its map of potentiality spreads out like a wave. A single electron particle fired at a barrier with two slits, if the particle is watched by a person or a detector, will pass through one slit of the barrier as a particle. If it is not observed, it will pass through both slits as a wave - showing a double wave interference pattern on a screen beyond the barrier. Various versions of this experiment have been tried with the intent of outsmarting the particle. All have failed, sometimes with the particle seeming to know beforehand what the strategy will be. This is not what your down to earth common-sense scientist likes to see.

Rehovot, Israel, February 26, 1998: One of the most bizarre premises of quantum theory, which has long fascinated philosophers and physicists alike, states that by the very act of watching, the observer affects the observed reality.

In a study reported in an issue of Nature (Vol. 391, pp. 871-874), physicists conducted a highly controlled experiment demonstrating how a beam of electrons is affected by the act of being observed. The experiment revealed that the greater the amount of 'watching,' the greater the observer's influence on what takes place.

Quantum Mechanics states that particles can also behave as waves. When behaving as waves, they can simultaneously pass through several openings in a barrier and then meet again at the other side of the barrier. This meeting is known as interference.

Strange as it may sound, interference can only occur when no one is watching. Once an observer begins to watch the particles going through the openings, the picture changes dramatically: if a particle can be seen going through one opening, then it's clear it didn't go through another. In other words, when under observation, electrons are being "forced" to behave like particles and not like waves. Thus, the mere act of observation affects the experimental findings.

To demonstrate this, Weizmann Institute researchers built a tiny device measuring less than one micron in size, which had a barrier with two openings. They then sent a current of electrons towards the barrier. The 'observer' in this experiment wasn't human. Institute scientists used for this purpose a tiny but sophisticated electronic detector that can spot passing electrons. This 'observer's' capacity to detect electrons could be altered by changing its electrical conductivity, or the strength of the current passing through it.

Apart from 'observing' or detecting the electrons, the detector had no effect on the current. Yet the scientists found that the very presence of the detector (observer) near one of the openings caused changes in the interference pattern of the electron waves passing through the openings of the barrier. In fact, this effect was dependent on the amount of the observation: when the 'observer's' capacity to detect electrons increased, in other words, when the level of the observation went up, the interference weakened; in contrast, when its capacity to detect electrons was reduced, in other words, when the observation slackened, the interference increased

Thus, by controlling the properties of the quantum observer (detector) the scientists managed to control the extent of its influence on the electrons' behaviour. The theoretical basis for this phenomenon was developed several years ago by a few physicists, including Dr. Adi Stern and Prof.Yoseph Imry of the Weizmann Institute of Science, together with Prof. Yakir Aharonov of Tel Aviv University.

Quantum Mechanics says all particles in nature whether photons, electrons, quarks, neutrons or other, behave in this wave/particle manner.

This experiment and many others cemented the fact that sub-atomic particles are affected by our looking. Philosophical physicist John A Wheeler states, "In some strange way the Universe is a participatory Universe. No theory of physics that deals only with physics will ever explain physics. I believe that as we go trying to understand the Universe, we are at the same time trying to understand man. The physical world is in some deep sense tied to the Human being." Wheelers thoughts along these lines were purely a kind of physical theorising, and he did not agree with the truth of the Universe created by God.

When the observer effect was first noticed early in Quantum Mechanics it caused much consternation. It undermined a basic of science: there is an objective world out there, separate from of us. If matters behaviour depends on how we look at it, what does "reality" really mean?

Could the simple answer to the quantum mysteries be: our preconceived idea about so called inanimate matter is wrong? Most of the response to the observer effect has been toward our consciousness seeming to effect matter. The Bible has it the other way around. It seems to attribute consciousness of a sort, to matter. The idea does not seem to have occurred to physicists. Yet that could be what they have found. It may have passed through the mind of some but speaking openly about such things would be hazardous to your career as a common-sense scientist. John A Wheeler took some flak for his philosophical ideas along those lines. Matter having a sort of consciousness is the best I can come up with to describe it. Examples from the Bible demonstrate my meaning. The Bible attributes characteristics to created matter. I realised I had not given much thought to these many passages in the Bible.

Uebert Angel draws our attention to Mark, where Jesus finding no figs on a fig tree - He 'answered' it saying, "Let no one eat fruit from you ever again." (Mark 11:14) By both answering the tree and using the word 'you', Jesus seems to be having a conversation with the tree. Now, don't blame me, I'm just passing on what's in the Bible!!

We thought it was the wind blowing the leaves in the trees but Isaiah 55:12 tells us the trees of the field clap their hands as the joyous person blessed by God passes by. How do they know the persons experience? And there is more: the mountains and the hills will break forth into singing before that person. Are we missing something about creation?

"The heavens declare the glory of God; and the firmament shows His handiwork. Day unto day utters speech. And night unto night reveals knowledge. There is no speech nor language where their voice is not heard. Their line has gone out through all the earth and their words to the end of the world." (Psalm19:1-4) We would not bat an eyelid if this passage stopped at 'The heavens declare the glory of God'. No problem, it's a metaphor. But it is not finished. The firmament speaks, it has a voice, and it uses words.

After Cain killed his brother Abel, God said to him, "What have you done? The voice of your brothers blood cries out to me from the ground." (Gen 4:10) Blood has a voice. Apostle Paul confirms that in Hebrews 12:24. "...to Jesus the mediator of the new covenant, and to the blood of sprinkling (Jesus blood) that speaks better things than that of Abel. Perhaps I do not need to 'plead the blood'. The blood of Jesus is always speaking for me in the ears of God the father.

"We know, Paul says, the whole of creation groans and travails waiting for the glorious liberty of the sons of God." Then he equates that groaning with our own groaning as we wait for the redemption of our bodies. (Romans 8:21-23) Verse 19 of that chapter says all of creation has an earnest expectation, eagerly awaiting the revealing of the sons of

God. We don't normally attribute those characteristics to lifeless matter (as we call it) The NIV translation commands us to 'preach the gospel to all creation.' Does that mean we are to tell the earth, the trees, the rivers, the lakes, and the animals, "not long now, because of Christs atonement, the sons of God will soon be revealed?"

I have heard many testimonies from people who spent time in heaven. They often say the flowers, and rivers and grass respond to us in heaven. That is not a fairy story. It has the testimony of many witnesses. Is that how it was on earth before the fall? Is that why the Bible attributes life like qualities to nature. Is nature longing to get back to its proper state again? Have the physicists stumbled on a residue of the proper form of created matter? Jesus declared He was introducing the Kingdom of heaven to the earth. He told us to pray, "Our Father who art in heaven, hallowed be thy name. Your Kingdom come your will be done on earth as it is in heaven." The beginning of restoration of the original unity between heaven and earth lost at the fall.

Though physicist John A Wheeler was not a Christian, his statements reveal someone beginning to get a fleeting glimpse of the truth of creation. Quote: "I think we are beginning to suspect that man is not a tiny cog that doesn't really make much difference to the running of the huge machine but rather that there is a much more intimate tie between man and the Universe than we heretofore suspected."

Quote:" Consider if the particles and their properties are not somehow related to making man possible. Man, the start of the analysis, man, the end of the analysis—because the physical world is, in some deep sense, tied to the human being."

Quote: "We live on an island surrounded by a sea of ignorance. As our island of knowledge grows, so does the shore of our ignorance."

Quote: "Is the very mechanism for the Universe to come into being meaningless or unworkable or both unless the Universe is guaranteed to produce life, consciousness and observer-ship somewhere and for some little time in its history-to-be?"

Quote: "No theory of physics that deals only with physics will ever explain physics. I believe that as we go on trying to understand the Universe, we are at the same time trying to understand man."

Quote:" The Universe does not exist "out there," independent of us. We are inescapably involved in bringing about that which appears to be happening. We are not only observers. We are participators. In some strange sense, this is a participatory Universe. Physics is no longer satisfied with insights only into particles, fields of force, into geometry, or even into time and space. Today we demand of physics some understanding of existence itself."

Wheeler did not accept Jesus' divinity but saw Him as a powerful example of the right way to live. Sadly, he seems not to have progressed from that belief in his lifetime. I'll add a comment here - the Bible is the best record of the life of Christ. One can barely get past one page of reading without being confronted with Christs divinity in all He said and did.

Albert Einstein was not impressed with many of the new findings of Quantum Mechanics. Mr Einstein disparaged quantum entanglement calling it "spooky action at a distance.". Of the observer effect he said, "the Moon does not exist only when we look at it!" I wonder if he considered, of the few billion people living on earth, the odds of at least one person looking at the moon at any moment in time are very high?! Joking again! Einstein maintained the belief until he died that a 'common sense' explanation would be found for the weirdness of Quantum Mechanics.

It does bring up another problem in Quantum Mechanics for which satisfactory answers have not yet been found. If all these enigmas are happening at the quantum level, why are they not happening at the macroscopic level? The problem is referred to as 'the gap'. To date, satisfactory answers have not been forthcoming.

"The Nature of the Physical World," is a book by Sir Arthur S. Eddington an eminent physicist of the early 1900's. The book is based on his lectures. He tackles the gap between the quantum field and reality as we experience it by referring to his 'two' tables.

"I have settled down to the task of writing these lectures and have drawn up my two chairs to my two tables. Two tables! Yes, there are duplicates of every object about me - two tables, two chairs, two of everything.

One of them has been familiar to me from earliest years. It is a commonplace object of that environment which I call the world. How shall I describe it? It has extension; it is comparatively permanent; it is coloured; above all it is substantial. By substantial I do not merely mean that it does not collapse when I lean upon it; I mean that it is constituted of "substance" and by that word I am trying to convey to you some conception of its intrinsic nature. It is a thing; not like space, which is a mere negation; nor like time, which is - Heaven knows what! it is the distinctive characteristic of a "thing" to have this substantiality, and I do not think substantiality can be described better than by saying that it is the kind of nature exemplified by an ordinary table. After all, if you are a plain common-sense man, not too much worried with scientific scruples, you will be confident that you understand the nature of an ordinary table.

Table Number two is my scientific table. It is a more recent acquaintance and I do not feel so familiar with it. It does not belong to the world previously mentioned - that world which spontaneously appears around me when I open my eyes, though how much of it is objective and how much is subjective I do not here consider. It is part of a world which in more devious ways has forced itself on my attention. My scientific table is mostly emptiness. Sparsely scattered in that emptiness are numerous electric charges rushing about with great speed; but their combined bulk amounts to less than a billionth of the bulk of the table itself. Notwithstanding its strange construction it turns out to be an entirely efficient table. It supports my writing paper as satisfactorily as Table number one; for when I lay the paper on it the little electric particles with their headlong speed keep on hitting the underside, so that the paper is maintained at a nearly steady level. If I lean upon this table I shall not go through; or, to be strictly accurate, the chance of my scientific elbow going through my scientific table is so excessively small that it can be neglected in practical life. Reviewing their properties one by one, there seems to be nothing to choose between the two tables for ordinary purposes.

There is nothing substantial about my second table. It is nearly all empty space - space pervaded, it is true, by fields of force, but these are assigned to the category of "influences," not of "things." Even in the minute part which is not empty we must not transfer the old notion of substance. In dissecting matter into electric charges, we have travelled far from that picture of it which first gave rise to the conception of substance, meaning experience. Whether we are studying a material object, a magnetic field, a geometrical figure, or a duration of time, our scientific information is summed up in measures; neither the apparatus of measurement nor the mode of using it suggests that there is anything essentially different in these problems..

I will not here stress further the non-substantiality of electrons, since it is scarcely necessary to the present line of thought. Conceive them as substantially as you will, there is a vast difference between my scientific table with its substance (if any) thinly scattered in specks in a region mostly empty and the table of everyday conception which we regard as the type of solid reality. It makes all the difference in the world whether the paper before me is poised as it were on a swarm of flies and sustained in a shuttlecock fashion by a series of tiny blows from the swarm underneath, or whether it is supported because there is substance below it, it being the intrinsic nature of substance to occupy space to the exclusion at least, but no difference to my practical task of writing on the paper.

I need not tell you that modern physics has by delicate test and remorseless logic assured me that my second scientific table is the only one which is really there! - wherever "there" may be. On the other hand, I need not tell you that modern physics will never succeed in exorcising that first table - strange compound of external nature, mental imagery, and inherited prejudice - which lies visible to my eyes and tangible to my grasp.

Yes, no doubt they are ultimately to be identified after some fashion. But the process by which the external world of physics is transformed into a world of familiar acquaintance in human consciousness is outside the scope of physics. And so, the world studied according to the methods of physics remains detached from the world familiar to consciousness, until after the physicist has fashioned his labours upon it. Provisionally, therefore, we regard the table which is the subject of physical research as altogether separate from the familiar table, without prejudging the question of their ultimate identification.

It is true that the whole scientific inquiry starts from the familiar and in the end, it must return to the familiar world. But the part of the journey over which the physicist has charge is in foreign territory."
Unquote

Quantum particles sometimes act like waves, spread out in space. They can slosh into each other and even back onto themselves. But if you poke at this wave-like object with certain instruments, or if the object interacts in specific ways with nearby particles, it loses its wavelike properties and starts acting like a discrete point—a particle. Physicists have observed atoms, electrons, and other minutiae transitioning between wave-like and particle-like states for decades.

But at what size do quantum effects no longer apply. How big can something be and still behave like both a particle and a wave? Efforts to answer that question have been hindered because the necessary experiments have been almost impossible to arrange.

Markus Arndt is a Professor of Quantum Nanophysics at the University of Vienna. He and his research team hold the record for observing quantum properties in large objects. They have observed wave like properties in molecules composed of 2000 atoms. This feat beats the previous record by two and a half times. The molecules were injected into a 5-meter-long tube. When the particles hit a target at the end, they didn't just land as randomly scattered points. Instead, they formed an interference pattern, a striped pattern of dark and light stripes that suggests waves colliding and combining with each other. They published the work in Nature Physics.

Of the difficulty Timothy Kovachy of North-Western University, who was not involved in the experiment says, "It's an extremely difficult experiment to pull off because quantum objects are delicate, transitioning suddenly from their wavelike state to their particle-like one via interactions with their environment. The larger the object, the more likely it is to knock into something, heat up, or even break apart, which triggers these transitions."

To maintain the molecules in a wave-like state, the team clears a narrow path for them through the tube, like police cordoning off a parade route. They keep the tube in a vacuum and prevent the entire instrument from wobbling even the slightest bit using a system of springs and brakes. It never moves more than 10 nanometres. The physicists then had to carefully control the molecules' speed, so they don't heat up too much.

One possibility physicists are exploring is that quantum mechanics might in fact apply at all scales. "You and I, while we sit and talk, do not feel quantum," says Arndt. We seem to have distinct outlines and do not crash and combine with each other like waves in a pond. "The question is, why does the world look so normal when quantum mechanics is so weird?"

By looking for wavelike behaviour in progressively larger objects, Arndt wants to understand how quantum mechanics transitions into the world we normally perceive. Many theories have been mooted but none have settled the issue. To perform the experiment, Arndt's team used a green laser to launch the molecules into the tube. The molecules absorbed the energy from the light to propel them forward. Then, the molecules passed through a sequence of metal grates containing thin, nanometres wide slits. The grates divide a single molecule into multiple waves traveling in different directions and recombine them in the end to form the interference pattern. It's a version of the previous mentioned double slit experiment.

They were careful to choose the optimal type of molecule for the experiment. They settled on a synthetic jumbo-sized object with the chemical formula, $C_{707}H_{260}F_{908}N_{16}S_{53}Zn_4$. Its structure was sturdy enough so that its peripheral atoms wouldn't fall off during launch. It also contains a core assortment of atoms called a porphyrin, which absorbs green light to act as the molecule's motor.

Now, Arndt's team plans to run this experiment for even more massive objects. They want to test whether they can observe wave-like properties in metal nanoparticles ten times heavier than their molecule. Researchers are working toward creating wave-like interference in objects even closer to the macroscopic realm. By doing these experiments, physicists hope to find the seam where the two places meet.

Quantum Darwinism is an idea put forward by physicist Wojciech Zurek at Los Alamos National Labs in New Mexico. It is the latest in theories attempting to explain the gap. Its main claim is that it explains the quantum-classical transition: why macroscopic physics obeys classical rules while the quantum world obeys the seemingly weird laws of quantum mechanics. As far as I can tell only a mathematician or physicist – these people of remarkably high intelligence – would understand the theory. A vague description can be offered in terms of the Darwinian part of the title. Only some of the quantum field is strong enough to break through to the reality realm. These become 'pointer states' that make up the solidity we take for granted. It has sparked interest among Zureks' colleagues, and teams are attempting to advance the theory in a few locations around the world.

TUNNELLING

Quantum tunnelling: The label that has stuck to this quantum phenomenon is misleading. When sub-atomic particles encounter a barrier they do not have enough energy to overcome, most either bounce/reflect off the barrier or bury themselves in it. For reasons that are not understood every so often one of the particles will dematerialise and appear on the other side of the barrier. The mystery of this can be illustrated by the picture of 100 balls rolling towards a steep hill. On reaching the hill, 99 of the balls get a certain distance up the hill then roll back. One of the balls inexplicably disappears and reappears on the far side of the hill. This is classic quantum weirdness.

The following testimony has nothing to do with quantum tunnelling as far as I know. It does have dematerialising and reappearing though. David Hogan is always in jeans and cowboy boots, even when he preaches. He is a proud redneck and sees no reason to change. David and his wife have been Missionaries to the Indian people in the jungles of Mexico for many years. It's a wild and woolly part of the world where police protection is days away. David tells of a time he responded to a request from a woman sick with cancer. She asked for prayer with laying on of hands. It would be a long trip through an area with a reputation for violence. David has a heart of love for the Mexican people. He climbed aboard his trail bike and took off. Not far from his destination as he passed through a village, a band of armed guerrillas came out of the trees. He was too fast for them and sped away. After doing all he could for the sick woman he was on the trail bike again. Jungle areas mean you must stick to established trails. The guerrillas were ready for him this time.

His choice was stop or be shot. The leader of the armed group walked up and wacked David on the head with the butt of his gun. David prepared himself to go down fighting. Just then a small man stepped out of the bush and called a halt to the intended violence saying, "this man helps our people," It's not clear even when David tells it, if the small man was an angel or a leader of the guerrillas. They let him go with the charge, "next time you come back tell us about your Jesus" David gunned the bike and left. Not much further on was a steep hill. In his elation at being free he forgot the steep drop just beyond the summit of the hill. He remembered as he and the bike flew out into space.

The next thing he was aware of was sitting on the bike, it was switched off, on the ground on the trail to his home. He started the bike and rode off. It had been an eventful day. *Unquote*

It turns out quantum tunnelling is a must have for our life on this earth. Quantum tunnelling answered the longstanding question why our Sun has been burning for so long without burning out and is expected to burn a lot longer. The Sun is a collection of gases mostly hydrogen and helium, held together by gravity. The nuclear fusion/burning in the Sun has been going for a long time and it will be generating heat and light for eons to come. The 'burning' in the Sun is created by hydrogen atoms fusing to make helium which causes a release of energy/nuclear reaction like the explosion of the atomic bomb. This creates the heat and light we get from the Sun. 600 million tons of hydrogen is fused into 596 million tons of helium per second in the Sun. What does quantum tunnelling have to do with this?

Hydrogen atoms have a positive electric charge. Hence, they repel each other. How then to get the fusing that's needed to create helium? Quantum tunnelling to the rescue! Every so often a hydrogen atom mysteriously crosses the double positive repel barrier and unites with its fellow atom. This happens enough to give us the heat and light we need from the Sun and causes the Sun to be a 'slow burn' ensuring its long life. The Sun does not need oxygen to burn. It is one big nuclear explosion.

An amazing demonstration of Gods handiwork. The Lord certainly gave us the simplified version in Genesis 1:16-18. "Then God made two great lights; the greater light to rule the day and the lesser light to rule the night. He made the stars also. God set them in the firmament of the heavens to give light on the earth, and to rule over the day and over the night, and to divide the light from the darkness. And God saw that it was good." A lot was going on when that happened. Especially since every star is a duplicate of our Sun! And it only took a day to get it going!! *Here is a trivia fact about the Sun. It moves through space in an enormous orbit that takes 260 million years. While we are out in space, another fact. Reliable estimates now say the Universe contains more than 50 billion galaxies with around 200 million stars in each.* God says, "I count the number of the stars; and call them all by name." (Psalm 147:4) He lets us in on some of those names in Job, "Can you bind the cluster of Pleiades, or loose the belt of Orion? Can you bring out Mazzaroth (Constellations) in its season? Or can you guide the great bear with its cubs?" (Job 38:31-32)

Quantum tunnelling also happens in the nucleus of heavier elements. An atom with a nucleus that is too big releases what is called an alpha particle made up of two protons and two neutrons. But the energy barrier of the nucleus the alpha particle must pass through is much greater than the energy of the alpha particle. How does it get out? Quantum tunnelling. It's the disappearing and appearing on the other side sleight of hand again. Another of the inexplicable conundrums of Quantum Mechanics.

Here is a more scientific description of the process: Quantum mechanics, however, allows the alpha particle to escape via quantum tunnelling. The quantum tunnelling theory of alpha decay, independently developed by George Gamow and Ronald Wilfred Gurney and Edward Condon in 1928, was hailed as a very striking confirmation of quantum theory. Essentially, the alpha particle escapes from the nucleus not by acquiring enough energy to pass over the wall confining it, but by tunnelling through the wall. Gurney and Condon made the following observation in their paper on it: It has hitherto been necessary to postulate some special arbitrary 'instability' of the nucleus; but in the following note it is pointed out that disintegration is a natural consequence of the laws of quantum mechanics without any special hypothesis. Much has been written of the explosive violence with which the α-particle is hurled from its place in the nucleus. But from the process pictured above, one would rather say that the α-particle almost slips away unnoticed. *Unquote*

EVIDENCE

E vidence is mounting quantum strangeness is a central player in many vital processes of life on earth.

Enzymes are protein molecules in cells which work as catalysts - speeding up chemical reactions in the body, but do not get used up themselves in the process. Almost all biochemical reactions in living things need enzymes. With an enzyme, reactions work much faster than they would without.

There are thousands of varieties of enzymes. Enzyme names usually end in –ase to show that they are enzymes. Examples include DNA polymerase. It reads an intact DNA strand and uses it as a template to make a new strand. Amylase found in saliva, breaks down starch molecules into smaller glucose and maltose molecules. Another is lipase which breaks down fats into smaller molecules.

Enzymes speed up chemical reactions so that processes that would take thousands of years happen in seconds inside living cells. We could not live without them. Just how they accelerate chemical reactions, often more than a trillion-fold, has long been an unanswered question. Experiments over the past few decades, however, have shown that enzymes make use of quantum tunnelling to accelerate biochemical reactions. The enzyme encourages electrons and protons to vanish from one position in a biomolecule and instantly materialise in another, without passing through the gap in between. Another instance of Quantum Mechanics undergirding life on earth.

In photosynthesis plants soak up some of the 1017 joules of solar energy that bathe Earth each second, harvesting as much as 95 percent of it from the light they absorb. The transformation of sunlight into carbohydrates takes place in one million billionths of a second, preventing much of that energy from dissipating as heat. How plants manage this almost instantaneous trick has been hidden. Now biophysicists at the University of California, Berkeley, have shown that plants use what will be a basic principle of quantum computing - the exploration of a multiplicity of different answers at the same time.

Plants all share a feature known as a photosynthetic reaction centre. Pigments and proteins found in the reaction centre help organisms perform the initial stage of energy conversion.

These pigment molecules, or chromophores, are responsible for absorbing the energy carried by incoming light. After a photon hits the cell, it excites one of the electrons inside the chromophore. Using the latest instrumentation scientists observed the initial step of the process and saw something no one had observed previously: a single photon appeared to excite different chromophores simultaneously.

"The behaviour we were able to see at very fast time scales implies a much more sophisticated mixing of electronic states," a spokesman said. "It shows us that high-level biological systems could be tapped into very fundamental physics in a way that didn't seem likely or even possible."

"The quantum effects observed in the course of the experiment hint that the natural light harvesting processes involved in photosynthesis may be more efficient than previously indicated by classical biophysics," said one scientist. "It leaves us wondering: how did mother nature (read God) create this incredibly elegant solution?"

Biophysicist Gregory Engel and his colleagues cooled a green sulfur bacterium - chlorobium tepidum, one of the oldest photosynthesizers on the planet, to minus 321 degrees Fahrenheit and then pulsed it with extremely short bursts of laser light. By manipulating these pulses, the researchers could track the flow of energy through the bacterium's photosynthetic system. "We always thought of it as hopping through the system, the same way that you or I might run through a maze of bushes,"

Engel explains. "But, instead of coming to an intersection and going left or right, it can actually go in both directions at once and explore many different paths most efficiently."

In other words, plants are employing the basic principles of quantum mechanics to transfer energy from chromophore (photosynthetic molecule) to chromophore until it reaches the so-called reaction centre where photosynthesis, as it is classically defined, takes place. The particles of energy are behaving like waves. "We see very strong evidence for a wavelike motion of energy through these photosynthetic complexes," Engel said," employing this process allows the near perfect efficiency of plants in harvesting energy from sunlight and is likely to be used by all of them. Lacking any navigational sense, most photon energy should hop aimlessly in the wrong direction, ending up in the metaphorical water." And yet, inside plants and bacteria that perform photosynthesis, nearly all packets of photon energy reach the reaction centre.

We have the simple version again in Genesis 1:3, "And God said, 'Let there be light'; and there was light." Notice the creation of light was Gods first declaration. Most of us realise light is essential for life on Earth. It had to come first. Quantum Mechanics has revealed a process that jumps to the top of the list of ways light is crucial to our existence.

We need to go down again to the micro world inside the atom. All matter, including our bodies, is composed of atoms. We have read the nucleus of the atom is positively charged while the electrons flying around the nucleus are negatively charged. Positives and negatives attract each other so we would expect the electrons to plunge into the nucleus causing everything including ourselves, to collapse into nothing.

But the catastrophe does not happen. It is prevented by light. Physicist Richard Feynman established that photons of light are constantly being exchanged between the nucleus and the electrons in an interaction that is enough to keep the electrons in place. Author Brian Clegg says, "This means each atom, each physical object, each of us – contains an absolute fireball of light, busily ensuring that matter stays intact. It's not visible – it doesn't get out of the tight world of the atom – but it's there inside us. We are truly creatures of light."

A chasm exists between physical light and spiritual light. I came upon a puzzle at Genesis 1:3-5. God creates light but it does not seem to be light from the Sun. The Sun is created on the fourth day. One is tempted to think this first creation is spiritual light. That explanation is defeated in verse five where God calls the light day and the darkness night – on the first day.

Leaving that mystery unsolved we move to Mathew 4:15,16 where the difference between physical and spiritual light is exemplified. "The land of Zebulon and the land of Naphtali, by way of the sea beyond Jordan. Galilee of the Gentiles: the people who sat in darkness have seen a great light. And upon those who sat in the region and shadow of death Light has dawned." Here is a region where the sun shines in its full strength, yet it is described as a land of shadows and darkness. The light that came is Jesus. The land was in darkness because it knew nothing of Him.

I (the author) had a miraculous experience of spiritual light many years ago while a new Christian. I had been working on a construction site and was aboard a small truck with other men, on our way home. An accident occurred at an intersection and my arm was slammed against the steel of the truck tray. I stumbled out of the truck suffering from shock. They lay me on a bed in an ambulance when it arrived. My first knowledge of injury to my arm was when the attendant lifted it and I saw a 3cm wide flow of blood under the skin from the injury area. At the time I was a financial supporter of the Oral Roberts Ministry. The attendant placed my arm under the blanket. Immediately I became aware of the anointing of the Oral Roberts Ministry, and though my arm was under the blanket I saw a ball of spiritual light around the injury area. When I lifted my arm from under the blanket the blood flow had gone. The anointing stayed with me and after lying on a gurney for fifteen minutes at the hospital, I left without telling anybody and walked home - around

10km. I visited Christian friends on the way and testified of what God had done. The residual effects were a swollen arm and a headache from the shock. Both were gone in a short time.

TIME

Time looms large in the Bibles proof of itself as the Word of God. In the form of prophecies fulfilled. 'Beyond Time and Space' is a book in which Chuck Missler discusses Bible prophecy. Many prophecies detail the life and death of Jesus as the Saviour of the world. Mathew 4:15,16 previously mentioned is fulfilment of the prophecy of Isaiah 9:1,2. Psalm 22 and Isaiah 53 give more information concerning Jesus death, burial and resurrection than does the New Testament. In each case the time between prophecy and fulfilment is many hundreds of years. The good lady Aiko Hormann, mentioned at the beginning of this book, received Christ after a workmate challenged her to study the fulfilment of prophecy in the Bible. In her own words, "As I became more intrigued with Bible prophecy fulfilled, I began to neglect my work as a scientist." That intrigue became a world-wide miracle ministry.

I must be honest - I find it difficult to understand Albert Einstein's special theory of relativity. I am willing to accept experts' interpretation of the theory. It makes time a physical entity as part of spacetime. Spacetime is bent by gravity in the vicinity of stars and planets. Apparently, time slows down as an object or person nears the speed of light. If that person were able to go faster than light they would begin to go back in time. Gerald Feinberg back in 1969 imagined a faster than light particle which he called a tachyon. The biggest benefactors of the hypothetical tachyon have been the science fiction writers. They ran with the concept and several best sellers were the result.

Cosmic rays are showers of particles that regularly collide with the top of Earths' atmosphere. At the collision with atomic particles at top of the atmosphere, they break into smaller particles which can be detected at ground level. Any tachyons among them would travel back in time and arrive before the other particles and the cosmic rays themselves. Hence, they would show up on detectors as blips before the rest arrive. Researchers are constantly looking for these blips. Some excitement ensued in 1973 when Australians Roger Clay and Philip Crouch found such blips. They sent the results to the publication 'Nature'. The details were published in 1974 and the press made much of it. Those results are no longer taken seriously because no other detector has ever had supporting blips. Nick Herbert, a physicist at Stanton, ranks the probability of the existence of tachyons slightly higher than the existence of Unicorns! Tachyons remain a hypothetical particle.

Another factor that changes time is time dilation. Einstein's theories predicted that time passes differently throughout the universe. This has been proven to be correct - clocks tick slower on the International Space Station (ISS) than they do here on Earth. It happens because time moves slower for objects that are near strong gravitational fields - such as Earth - than for objects further from the fields - like the ISS.

For a long time, Russian cosmonaut Sergei Vasilyevich Avdeyev held the record for time dilation experienced by a human being. In his 747 days aboard the Mir Space Station, he went approximately 27,360 km/h, orbited the earth 11,968 times traveling about 515,000,000 kilometres and aged roughly 0.02 seconds (20 milliseconds) less than an Earthbound person would have. Considerably more than any other human being at the time. He time travelled one forty-eighth of a second to the future. Since then Sergei Krikalev and Gennady Padalka have extended the record which at present is 879 days in space in 2015. A common misconception is that the Apollo astronauts hold the record - they did go faster than Avdeyev, but they were only in space for a few days. Those statistics are the progress to date regarding science and time travel.

I read of a scientist vehemently disagreeing with the idea our futures are fixed. It got me thinking about the book of Revelation and the future revealed therein. God sees all of time from His vantage point in eternity. Galatians 4:4 tells us God in His wisdom sent Jesus at a certain point in time of His choosing. Responding to a question from the twelve Jesus said, "It is not for you to know times or seasons which the Father has put in His own authority." (Acts 1:7) God called John up into the eternal realm to view time (the future) as God sees it. (Rev 4:1)

Do the contents of the book of Revelation mean our futures are fixed? No. Though we cannot escape time we have freedom of choice within it. It is just that God can see the results of our free choice in His vista of time. He then works His purposes into time from that foreknowledge. The same applies to our choice to accept Christ. Are some chosen for salvation and others not? No. God sees through all time those who will accept Jesus and those who won't. That is why the New Testament calls us "Chosen in Him before the foundation of the world." (Ephes 1:4) It does not say chosen, rather, chosen in Him. It could not be any other way. We are told whosoever will, may come to Christ. (John 3:16) Jesus made it clear in the Great Commission, salvation is a choice. (Mark 16:15,16)

John was shown the future, but Bruce D Allen was taken back in time before he had a chance to think about it. He tells it in his book 'Foundations of Glory'. "Let me share with you a story, I didn't share this for a long time because I had no paradigm for it. At the time it was something totally outside my grid. So, to speak. I was in my study praying one day and minding my own business when suddenly, I was literally moved by the Spirit of God, and I found myself in Antwerp, in the Middle Ages. It was pouring down rain: the streets were muddy and hard to traverse due to the mud. A wagon came slogging through the mud, and there was a little boy sitting there in the mud weeping. I went over to him and I said, 'What's the matter?' He said, 'My mommy is dying, and they won't let me see her.' I asked why. 'There is a plague, he replied.' I said, 'Take me to her.' And in his face, I saw a little spark of hope. He grabbed my hand and he led me through the squalor to a home, and we opened the door. There was a bed in the middle of the floor. A man was sitting there, the woman's husband, and someone else off in the corner. But I had no immediate revelation of who he was. I think it was an angel.

The man cried out, 'No! Don't come in! Plague!' I said it was OK. I entered the house and stood at the foot of the bed, and I looked at this woman. She was aware. I began to preach the Gospel to them. Then I asked if they would like to accept Jesus, and they all said yes! So, this family accepted the Lord as their Saviour, and then I prayed for the woman's healing and she came out of the bed totally healed! And then I was back in my office."

"The whole experience was a huge learning curve. 'What was that Lord?' I asked Him. The Lord said, 'Google it.' (He found bubonic plague had started in the Jewish quarter in Antwerp in the Middles Ages. Thousands were killed. They evicted the Jews because they blamed them for the plague.)

He continues. "I told the Lord, 'I'll never share this! Nobody would ever believe this! I don't even understand this myself. How can I have travelled back in time?' He said, 'Don't limit me. Before the foundation of the world, written into the curriculum of history was this encounter I brought you back to effect. You have just reached the point in your life when this encounter was enacted, but they had it way back then.' And I didn't really understand."

"I still did not want to share it. I felt that way until I talked with a friend who said, 'Yes, that has also happened to me, but I have never shared it.' I jokingly told him I couldn't imagine why!" *Unquote*

Science magazines often proclaim faster than light speeds, (superluminal) have been attained. In almost every case a closer reading finds particle entanglement and particle tunnelling are the topics under discussion

Joshua Mills of NewWineInternational.org believes as Christians, time should be our servant, not the other way around. His book 'Time and Eternity' records what God has been saying to him regarding time. It is a record of events in his meetings and personal experiences. At one meeting he received a word of knowledge God was taking somebody thirty years back in time. A woman present had been injured in a car accident thirty years previously and was still suffering ill effects. God did a miracle returning her to the physical condition she had been in before the accident

He tells of a personal experience with time reversal at his home in 2011. "Last year something unusual happened as I was downstairs working in my tool shed. I had the door open to our staircase because Janet Angela (his wife) had been going up and down the stairs. Suddenly I heard Janet Angela screaming. I looked out the door and there she was lying on the floor all hunched up. She had fallen down the stairs and hurt herself badly."

"A holy unction of the Spirit of God rose up within me and I declared, 'In the name of Jesus I take you back 15 minutes.' Do you know what happened? Janet Angela looked at me and said, 'What just happened?' I replied, 'You just fell down those stairs.' She told me that the last thing she remembered was being at the top of the stairs, and the next moment sitting at the bottom of the stairs completely unharmed. She didn't remember falling. Not one thing was wrong with her. Now if you knew Janet Angela you would understand that she bruises very easily. After I declared, 'I take you back 15 minutes,' there was not one bruise, not one scratch – nothing was wrong. It was an absolute miracle."
Unquote

NEXT

The most talked about next steps in Quantum Mechanics are quantum cryptography and quantum computers. The following is a quote regarding **quantum cryptography:**

Traditional cryptography works using keys: a sender uses one key to encode information, and a recipient uses another to decode the message. However, it's difficult to remove the risk of an eavesdropper, and keys can be compromised. This can be fixed using potentially unbreakable quantum key distribution (QKD). In QKD, information about the key is sent via photons that have been randomly polarized. This restricts the photon so that it vibrates in only one plane (the meaning of polarization) for example, up and down, or left to right. The recipient can use polarized filters to decipher the key and then use a chosen algorithm to securely encrypt a message. The secret data still gets sent over normal communication channels, but no one can decode the message unless they have the exact quantum key. That's tricky, because quantum rules dictate that "reading" the polarized photons will always change their states, and any attempt at eavesdropping will alert the communicators to a security breach.

Today companies such as BBN Technologies, Toshiba and ID Quantique, use QKD to design ultra-secure networks. In 2007 Switzerland tried out an ID Quantique product to provide a tamper-proof voting system during an election. And the first bank transfer using entangled QKD went ahead in Austria in 2004. This system promises to be highly secure, because if the photons are entangled, any changes to their quantum states made by interlopers would be immediately apparent to anyone monitoring the key-bearing particles. But this system doesn't yet work over large distances. So far, entangled photons have been transmitted over a maximum distance of about 88 miles. *Unquote*

A quote regarding **quantum computers:** A standard computer encodes information as a string of binary digits, or bits. Quantum computers supercharge processing power because they use quantum bits, or qubits, which exist in a superposition of states – (super position is a quantum mechanics term referring to the general indeterminate state of quanta – eg. photon - before it is measured.) until they are measured, qubits can be both "1" and "0" at the same time.

IBM and Google have recently produced the first superfast quantum computers. IBM allows people outside the company to run experiments on the quantum computer and offers paid quantum cloud services. The field is still in development, but there have been steps in the right direction. In 2011, D-Wave Systems revealed the D-Wave One - a 128-qubit processor, followed a year later by the 512-qubit D-Wave Two. The company says these are the world's first commercially available quantum computers. However, this claim has been met with scepticism, in part because it's still unclear whether D-Wave's qubits are entangled. Studies released in May found evidence of entanglement but only in a small subset of the computer's qubits. There is also uncertainty over whether the chips display any reliable quantum speed up. Still, NASA and Google have teamed up to form the Quantum Artificial Intelligence Lab based on a D-Wave Two. And scientists at the University of Bristol last year hooked up one of their traditional quantum chips to the Internet so anyone with a web browser can learn quantum coding.

Ordinary computers run calculations using trillions of switches built into microchips. Those switches are either "on" or "off." A quantum computer, however, uses atoms or subatomic particles for its calculations. Because such a particle can be more than one thing at the same time - at least until it is measured - it may be "on" or "off" or somewhere in between. That means quantum computers can run many calculations at the same time. They have the potential to be thousands of times faster than today's fastest machines. *Unquote*

Quantum dots: Wouldn't you know it! We are now making our own atoms! That is not entirely true. Scientists are creating artificial atoms using the mysterious quantum processes God has put into nature. What does God think about that? Obviously, I don't know the answer to that, but it does still seem to come under Gods original instruction to Adam to subdue the earth. (Gen 1:28) And the intention behind what is still a very new science, is to benefit mankind. Quantum dots were discovered by accident in 1978. They occur and are now created in semi-conductor substances. Conductor substances allow a free flow of electricity (silver is the best) Resistors stop electricity in its' tracks, (rubber, plastic) and semi-conductors allow the flow of electricity within narrow energy bands. (silicon)

Doped (atoms of gases added to the silicon to aid semi-conducting) silicon when layered in certain ways traps electrons in one, (quantum wells) two (quantum wires) or three (quantum dots) dimensional spaces. Electrons trapped this way begin to act as they would in an atom. They organise themselves and behave precisely as they would in an atom, though no atom nucleus is present. Quantum dots exist just below the surface of the silicon and remain separate from it. They are maintained by applied electrical charges and can be adjusted without consequence to the silicon. Which atom quantum dots resemble depends on the number of electrons present. They can be adjusted by electric charge to resemble any atom on the periodic table. Hence the term artificial atoms. The people who create these pseudo products have started their own periodic table, naming the atoms after researchers.

Quantum dots ability to absorb and emit light have proved beneficial in testing for and treatment of cancer. In the book 'Bio applications of nanoparticles' by Warren W Chan - researcher Xiaohu Gao explains: "By adjusting their size and composition Quantum dots can be prepared to emit fluorescent light from the ultra-violet throughout the visible and into the infra-red spectra. Important for use as biological probes, quantum dots can absorb and emit light very efficiently allowing highly sensitive detection relative to conventionally used organic dyes and fluorescent proteins. They are well suited to monitoring biological systems for long periods of time, which is important for developing robust sensors for cancer assay (testing) and for invivo imaging. (testing) Thus increasing detection speed and lowering cost." *unquote*

SEARCHING

M any of the early pioneers of Quantum Mechanics were Jewish and it is a wonder to me it did not occur to them to include the Bible in their investigations. A knowledge of the Torah, at the least, would have been an inescapable part of their early family life. The idea science and God don't mix seems to be standard fare in most of the scientific fraternity. The search for the 'theory of everything' is a phrase found often in any reading about Physics or Quantum Mechanics. Stephen Hawking expressed it this way. "If we do discover a complete theory, it should in time be understandable in broad principle by everyone, not just a few scientists. Then we shall all be able to take part in the discussion of why it is that we and the Universe exist. If we had an answer to that it would be the ultimate triumph of human reason - for then we would know the mind of God."

Another commentator says, ..." In that sense, everything since the Big Bang has been one giant quantum experiment, in which all the particles in the Universe, including those we think of as making up the Earth and our own bodies, are involved. But if theory tells us we're among the sets of particles involved in a giant quantum experiment, the position I've just outlined tells us we can't justify any statement about what has happened or is happening until the experiment is over. Only at the end, when we might perhaps imagine some technologically advanced alien experimenters in the future looking at the final state of the universe, can any meaningful statement be made."

"Of course, this final observation will never happen. By definition, no one is sitting outside the Universe waiting to observe the outcome at the end of time. And even if the idea of observers waiting outside the Universe made sense – which it doesn't – on this view their final observations still wouldn't allow them to say anything about what happened between the Big Bang and the end of time. We end up concluding that quantum theory doesn't allow us to justify making any scientific statement at all about the past, present or future."

A physics publication I looked at recently, alleged they would answer three questions. Why is there something rather than nothing? Why do we exist? Why this particular set of laws and not some others? "We claim, they said, it is possible to answer these questions purely within the realm of science and without invoking any divine being." The questions were not answered. The reader had to be satisfied with probability, possibility and 'if our theories are correct'. Science alone will never answer those questions. Science reveals the works of God. "For since the creation of the world His (God) invisible attributes are clearly seen, being understood by the things that are made, even His eternal power and Godhead, so that they are without excuse." (Romans 1:20)

A metaphor comes to mind of a driver and passenger travelling between cities. The driver pulls over admitting they are lost, then discovers the map is not in the glove box. The passenger reminds the driver the car is fitted with GPS technology, which provides not only a map but a guiding voice. Having no knowledge of the technology the driver thinks the passenger is trying to make a fool of him and refuses to listen. They drive on hoping for the best.

The above statements are descriptive of the lost condition of humanity. Jesus said He came "To seek and to save those who are lost." (Luke 19:10) The Bible is the GPS God has given us. The mind of God is revealed in its pages. It comes with His voice if we welcome a relationship with Him through Jesus Christ. The three questions are answered in John 3:16, "For God so loved the world…" Love is why there is something rather than nothing. Love is why we exist. Love is the why - of this particular set of laws.

Starr Daily was a gangster who worked for Al Capone. Jesus finally got through to him in prison during a stretch in solitary confinement. He wrote these words. *"We swim in an infinite ocean of love. To become increasingly conscious of our oneness with love, is the mark of intelligent self-interest. To this end, we do not labour and strain in our search for love. It is above, beneath and about us. It is seeking us."*

I want to finish this book with the most extraordinary words you will ever read about the material realm and matter. (Mathew 27:50-54) **"And Jesus cried out again with a loud voice and yielded up His spirit. Then, behold, the veil of the temple was torn in two from top to bottom: and the earth quaked and the rocks were split, and the graves were opened: and many bodies of the saints who had fallen asleep were raised; and coming out of the graves after His resurrection, they went into the holy city and appeared to many. So, when the centurion and those with him, who were guarding Jesus, saw the earthquake and the things that had happened, they feared greatly, saying, "Truly this was the Son of God."**

.

.

www.ingramcontent.com/pod-product-compliance
Lightning Source LLC
Chambersburg PA
CBHW030018190526

45157CB00016B/3118